Mazes

Ancient & Modern

Robert Field

Tarquin Publications

Conimbriga, near Coimbra, Portugal

I wish to thank all those who have helped me with this book and in particular: Jeff Saward of Caerdroia*, Adrian Fisher, Randoll Coate, Maggie Berkowitz, Elaine M. Goodwin, Stephen Cosh, David Neale and Jean Aston. I am also indebted to many previous writers on the subject of mazes and labyrinths - a bibliography will be found on the inside of the back cover. All photographs and drawings are mine except for the drawings on page 31 by Stephen Cosh and on page 60 by Adrian Fisher and the photographs on page 20 by Peter Ostrolenk, on page 32 by Dr. Iain Kendell, on page 38 (bottom) and page 64 by Maggie Berkowitz, on pages 44, 58, 59, 60, 61 by Adrian Fisher and on pages 56, 57 by Elaine M. Goodwin. I thank them all for allowing me to use them.

* Caerdroia is the journal of a mazes and labyrinths research group which was founded in 1980. Membership is open to all who are fascinated by mazes and the journal is published annually. The current edition and back issues of the journal are available from Caerdroia, 53 Thundersley Grove, Thundersley, Essex SS7 3EB. In addition, an information and reprint service is available to researchers, enthusiasts, maze visitors and maze builders.

Tarquin Publications
Stradbroke
Diss
Norfolk IP21 5JP
England

Labyrinths or Mazes?
One choice or many

Although the title of this book is 'Mazes, Ancient & Modern', the word 'labyrinth' is also used and perhaps the best starting point is to distinguish between them. Strictly speaking, the word 'labyrinth' should only be applied to two designs. The first is the 'Classical Labyrinth' and the second is the 'Medieval Christian Labyrinth' based on the famous example at Chartres in France. Indeed, it is often referred to as the 'Chartres' design. An essential property of a labyrinth is that there is only one pathway to the middle and that no choices are offered. Such designs are called 'unicursal' or 'unicursive'. It is easy to see the association with Christianity and the idea of a single pathway to salvation and it was sometimes the custom for pilgrims in cathedrals and at religious sites to follow such pathways on their knees to symbolise such a commitment.

The word 'maze' is used where at various points you are given a choice of pathways. In such designs, it is possible to get lost, both in trying to find the centre and in trying to get out again. The most famous example of this kind of maze, described in a most amusing fashion in 'Three Men and a Boat' by Jerome K. Jerome, is the hedge maze at Hampton Court near London. Such designs are called 'multicursal' or 'multicursive' and are seen essentially as puzzles or intellectual challenges. Some designs lend themselves to the discovery of simple rules which can be applied to decide which way to turn when offered a choice of pathways. It is the maze, not the labyrinth, which has been greatly developed into a pen and paper puzzle form. In this guise, the participant is presented with a different set of problems. Far from being surrounded by dense yew or conifer hedges where it is very hard to see far at all, he or she can see the whole maze as if from above. Designers of such puzzle mazes make certain that the path which does lead to the centre is well obscured by dead ends and repetitive loops.

It is rather ironic that the best known labyrinth of antiquity, that of the Minotaur at Knossos in Crete, could scarcely have been a labyrinth at all. If there was only a single route in and out, there would have been no need of a ball of thread! Perhaps this is why in common speech the words labyrinth and maze are interchangeable. In this book we shall try to use the words in their formal sense but not take the matter too seriously. There is a lot of fun and creative imagination in maze design and great interest in visiting mazes and trying to solve them. From a serious and pious art form the maze has developed into a fascinating branch of the leisure industry. I hope that you derive as much pleasure from them as I have done in researching, photographing and drawing the mazes for this book.

Robert Field

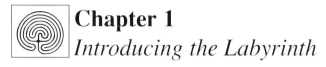

Chapter 1
Introducing the Labyrinth

The Story of Theseus and the Minotaur

Theseus was the son of King Aegeus of Athens and Aethra, daughter of the King of Troezen. Aegeus left Troezen before the birth of his son but, before leaving, buried his sandals and sword under a rock. He told Aethra that when the child was old enough to lift the rock and recover the sword and sandals he was to bring them to Athens and claim his birth-right. This Theseus did after many adventures on the way.

Soon after his arrival in Athen, the time came for the tribute to be sent from Athens to King Minos in Crete. Every nine years King Minos demanded seven youths and seven maidens be sent from Athens to him in Knossos. They were then sent into the labyrinth a building designed and built by Daedalus, the King's architect and craftsman, to house the Minotaur. This creature, half man and half bull, had the head of a bull on a human body. The youths and maidens were destined to be devoured by him. Theseus was so moved by the sorrow of the parents who had to send their children to Crete that he volunteered to go himself and be one of the victims if he was unable to kill the Minotaur.

When Theseus and his companions arrived in Crete they were taken to the palace at Knossos and brought before King Minos. His daughter Ariadne sat at his side and instantly fell in love with Theseus. She promised to help him to kill the Minotaur if he would marry her and take her to Athens. She then helped him to enter the labyrinth and gave him a ball of red thread and a sword. The plan was successful and he was able to kill the Minotaur and return safely to the entrance.

As with so many good stories there are sub-plots of betrayal and intrigue and the intervention of the Gods. Whatever its historical accuracy it is undoubtedly true that the legend of Theseus and the Minotaur is a significant part of our heritage. It has also contributed greatly to the development of the idea of the maze.

Bardo Museum, Tunis

Virtually a thousand years before the story of Theseus and the Minotaur was written down, the Minoan civilisation suddenly came to an end, probably as the result of a violent earthquake and a tidal wave. Over succeeding generations knowledge about this civilisation disappeared from the human memory and it was thought that the stories were just stories. Then, in 1922, the ruins of the palace of Knossos were discovered and excavated by Sir Arthur Evans. The excavation revealed an enormous warren of rooms and corridors and some magnificent frescos including the one above.

We know from the Greek historian Herodotus of a labyrinth built by Amenemhet III around 2000 B.C. in the Fayum district of Egypt. This building seems to have covered a great area and consisted of twelve courtyards and a vast number of rooms. Its exact purpose is not known but appears to have been a funerary monument with several uses. It is possible that such labyrinth buildings were built as tombs or temples for the dead kings. There was much contact and exchange of ideas and skills between Egypt and Crete and possibly the most famous of all ancient labyrinths, that at Knossos in Crete, was based on an Egyptian design of this sort.

However, later coins from Knossos show a typical Classical labyrinth with a single pathway.

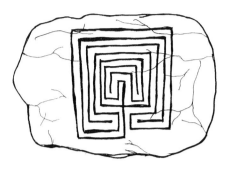

The palace of Pylos was an important centre in the Mycenean period of Greek history and the clay tablet illustrated above was preserved in a fire which destroyed the palace in about 1200 B.C. It is the earliest example of what is known as the 'Classical' labyrinth. It has a single entrance and the pathway to the centre passes through every part of it.

On the right is a rounded-off mirror image version. This is called a seven-ring labyrinth because a line from the centre to the outside along the line of rough symmetry crosses seven pathways.

(a)

(b)

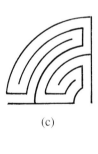

(c)

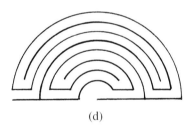

(d)

As to the origination of the actual design, Jeff Saward shows in his book 'Ancient Labyrinths of the World', how the classical meander pattern, used to decorate all manner of things from pottery to mosaics, could be extended into such a labyrinth by rotating the pattern in a systematic way, each time adjusting the lengths of the walls.

(e)

Drawing the Classical Labyrinth Design

If you would like to draw a Classical labyrinth, the best way is to start with a cross based on a square.

Defensive walls

The Iranian geographer Al-Biruni wrote a book around 1045 in which there is a drawing of the fortress of Lanka, shown right. This seemingly shows its defensive walls in the form of a classical labyrinth.

A complicated lay-out for a city offers many advantages to defenders. They know all the twists and turns but attackers do not know them and are placed at a disadvantage. However, it seems unlikely that anyone would use a standard design for such a purpose. It remains rather a mystery.

The Cities of Jericho and Troy
These two names often seem to be associated with mazes and the illustration above shows a diagram of the walls of Jericho. It was taken from a ninth century illustrated manuscript and suggests a labyrinth, not too dissimilar from the classical labyrinth pattern. In Finland there are several stone labyrinths which have names associated with Jericho. This is not too surprising as the biblical story of Joshua marching the Israelite army seven times around the city and then "The walls came tumbling down" is very well known. It does not seem at all unreasonable that this is why such a name might be chosen.

Troy is also a city which was famous for its great walls and the story of the wooden horse is similarly widely known. In addition, excavation of the site of the City of Troy has revealed many cities, each being built on the ruins of earlier ones. Perhaps, as in the case of Knossos, the idea of a labyrinth at Troy took strength from the size and complexity of these ruins. Whatever the reason turf mazes have been given names such as 'Troy Town', 'The Walls of Troy' and 'The City of Troy'. Another example is the word 'Caerdroia'. It is Welsh and is a nice play of words. When translated into English, it can either mean 'The Castle of Troy' or 'The Castle of Turnings'. A Welsh history book of 1740 refers to a tradition of shepherds cutting turf mazes for entertainment while watching their flocks. The name is now used by the British Maze and Labyrinth Society. (See page 2)

Hill Forts
Hill forts in Britain such as Maiden Castle and Tre'r Ceiri also had defended entrances which forced the attackers into single-file between high ramparts in a pathway that twisted back on itself. Although such maze-like entrances and indeed the overall ground plan may appear to suggest a labyrinth, it seems much more likely that this is just a coincidence. The purpose was simply to find the best method of defence in turbulent times.

The Medieval Christian Labyrinth, also know as the Chartres Design

This is the second of the labyrinth designs and perhaps the most famous example is the one that was laid in 1202 in Chartres Cathedral in France. It is positioned in the nave in such a way that, if the west wall was folded down on the floor, the magnificent rose window would cover the labyrinth.

There is only one entrance to the labyrinth and the design is divided roughly symmetrically into quarters. The red path shows the route from the entrance to the centre. Observe that there are no sections of pathway that are not traversed, an essential feature of any labyrinth. At the centre there used to be a brass plate incised with an illustration of Theseus and the Minotaur. However, it was removed and melted down during the Napoleonic wars and was never replaced.

The Chartres Design has been much copied although it is frequently modified to a greater or lesser extent. In this example on the right there are changes due to the centre being reduced in size. However, it remains an eleven-ring labyrinth. To count the rings of a labyrinth and so determine its type, start at the centre and go outwards on one of the four diagonal lines of symmetry. Count the number of pathways that you cross until you reach the outside. Choose a line of symmetry that crosses all the pathways, not one that misses any of the loops.

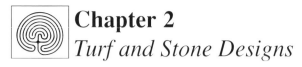

Chapter 2
Turf and Stone Designs

Turf Mazes

Turf is a dense mass of matted roots of grasses, other grassland plants and soil and if the plant cover is kept short by mowing or grazing, especially by sheep and rabbits, it will grow to form a sharply defined layer. This layer can be cut with metal tools and peeled off leaving a clean bare soil surface which is a few inches below the general level. This technique offers the possibility of sculpting shapes and pathways with sharp outlines and it is not surprising that it has been used to create mazes and labyrinths. It is thought that in Britain there were once more than two hundred such mazes although only eight have survived until today. Turf mazes seem to be a feature principally in Northern and Germanic countries, perhaps simply because this is where the climate and soil types allow the formation of good turf. Where the soil layer is thin, the turf layer can occupy all of it and so when it is removed, what remains is the underlying chalk, sand or gravel. In such situations sharp outlines can be maintained with minimal maintenance, whereas on rich deep soils the cut areas will quickly regrow and obscure them.

Cutters of a turf maze have two fundamental choices. Either the pathway can be on the turf which remains or it can be on the surface which is cut. Both kinds exist and of the eight which have survived six are of the first type and two are of the second. The designs of all of them are, or are based on, either the medieval 'Chartres' design or on the Classical labyrinth. All of them have a single pathway to the centre and out again and so are strictly labyrinths, not mazes. The reason why they were cut has been lost in the passing of time, but stories abound and they are given in the following pages. The most likely explanation is that they are linked either with a religious settlement or with ancient fertility rites. Let us start with the plans of four mazes which no longer exist.

1. Pimperne, near Blandford in Dorset

Writing in 1686 John Aubrey records that the maze ' ... was much used by the young people on Holydaies and by ye School-boies'. It was of a unique triangular design which we know from a drawing by J. Bastard made in 1758 and used in John Hutchins 'The History and Antiquities of the County of Dorset' written in 1774. The maze was destroyed by ploughing in 1730.

2. 'Robin Hood's Race' or 'The Shepherd's Race' at Sneinton, Nottinghamshire

This maze had four corner bastions and resembles the one which still exists at Saffron Walden. It is also unusual in that it seems to break the general rule and offers a choice of pathways just after the entrance. The circular bastions had heraldic crosses within them which were known as 'cross-crosslet fitchy'. This maze was destroyed by ploughing in 1797 after the enclosure of the common. Although its origins are unknown it is likely that it had a connection with a nearby monastic settlement at St. Anne's Chapel.

3. 'The Walls of Troy' between Paul and Marfleet, Holderness, Yorkshire

This is a twelve-sided version of the medieval labyrinth. Its plan was recorded in 1815 but it is not known when it disappeared. A circular version still survives on the opposite bank of the Humber at Alkborough (See page 18).

4. 'Walls of Troy' Rockliffe Marsh, Cumbria

This design could have been derived from the classical labyrinth or, it is suggested, from a simple spiral. Trace the pathway to identify where it changes from being generally anticlockwise to being generally clockwise.

Turf and Stone Designs

The eight surviving turf mazes in Britain now follow.

Mizmaze, St. Catherine's Hill, Winchester

1. Mizmaze, St. Catherine's Hill, Winchester, Hampshire

The overall outline of the maze is square, but its development from the medieval Christian design is indisputable. Its pathway is rather irregular and is cut between the turf banks. It has been 'straightened out' below to give a more clinical version. This maze lies on the top of St. Catherine's Hill, to the south of Winchester, and is on a site which has long been regarded as a hallowed spot. There was an Iron Age fort there and in around 1080 the Chapel of St. Catherine was built within its ramparts. The chapel was destroyed during the period of the dissolution of the monasteries around 1539 and was never rebuilt. However it seems unlikely that the maze was in any way connected with this chapel. Local tradition says that it was made by a schoolboy from Winchester College to while away a period of detention in the seventeenth century. The first reference to this maze is a plan by J. Nowell dating from 1710.

The design straightened out

The design in circular form

Town Maze, Saffron Walden, Essex

2. Town Maze, Saffron Walden, Essex

This maze is situated on the east side of the common at Saffron Walden and is the largest surviving turf maze in Britain. Like the Mizmaze at Winchester the pathway lies between the turf banks. It is a seventeen-ring medieval Christian design with the addition of four corner bastions sticking out like ears. At the centre is a large domed mound on which an ash tree grew until November 5th 1823 when it was burnt down during the Guy Fawkes celebrations. The first reference to this maze is dated 1699 and it records a payment of fifteen shillings for the re-cutting of the maze. Since that date it has been re-cut many times, including 1828, 1841, 1859 and 1887. In 1911 the pathway was laid with bricks but these deteriorated and in 1979 it was once again re-cut and the brick path was completely re-laid.

During the eighteenth century it is recorded that the young men of the town had a set of rules connected with the walking of the maze which resulted in the winning of a wager measured in gallons of beer. Also that a young girl stood at the centre while boys tried one at a time to race the maze in record time without stumbling.

3. City of Troy, Dalby, North Yorkshire

This is the smallest of the British mazes and it is a seven-ring Classical design. It is surrounded by a fence on a small road between Dalby and Terrington in a spectacular setting high on the Howardian Hills. York is a dozen miles or so to the south. Local tradition says that it is unlucky to run the maze more than nine times.

City of Troy, Dalby, North Yorkshire

4. Troy Town, Somerton, Oxfordshire

This is a large fifteen-ring Classical design maze. It is privately owned and is not accessible to the public without prior permission. It is curious that both of the surviving Classical design turf mazes in Britain have names associated with Troy.

Hilton, Cambridgeshire

5. Hilton Maze, Hilton, Cambridgeshire

This maze is found on the village green within the village of Hilton. In its present
form, it is of a nine-ring medieval Christian design. The pathway is of turf with gravel
beds in between. It took this form when it was re-cut in 1967 by Philip Dickinson. In
1988 it was again re-cut and the then existing banks, which helped runners to turn more
swiftly, were removed. It is presumed that the original design was of the eleven-ring
medieval Christian design and that the middle two rings disappeared when the monument was erected in the middle. On the pillar it states "William Sparrow, born 1641,
died at the age of 88, formed these circuits in 1660". It has been suggested that
William Sparrow at the age of nineteen possibly re-cut an existing maze which had
become overgrown during the time of the Commonwealth (1649-59). In that period
there was active suppression of 'the relics of vile heathenism' which included maypoles
and mazes and 1660 was the year of the restoration of Charles II as King.

Hilton Maze is an example of the fragility of turf mazes and exemplifies the importance
of re-cutting and regular maintenance following the plan of the original design. In 1854
'The History, Gazetteer and Directory of the County of Huntingdonshire' states that the
maze was composed of 'small pebbles on the ground'. J. Hutchins in his 'History of
Antiquities of the County of Dorset' of 1774, when discussing the maze at Pimperne,
mentions the mazes at Sneinton and Hilton. Of the latter he says 'Another very curious
example occurs on the village green in the parish of Hilton, Hunts., which instead of
being composed of the usual banks of turf, has its intricate design laid in a neat paving
of pebbles'. Whatever its origins it is now extremely well-kept.

Mizmaze, Breamore, Hampshire

6. Mizmaze, Breamore, Hampshire

This maze is in a spectacular setting on Breamore Down in the middle of a wood on the top of a hill known as Mizmaze Hill. It is reached either through the grounds of Breamore House or from the A338 Salisbury Road. It is an eleven-ring medieval Christian design with a circular mound in the middle. Local tradition associates the maze with the Priory of St. Michael which was situated on the River Avon in Breamore. Monks 'walked' the maze on their knees as a form of penance. It is also said that a man could run to nearby Gallows Hill and back in the time it took another to run the maze.

In 1783 Sir Edward Hulse ordered the maze to be re-cut because it had become over-grown and ever since it has been maintained by the Breamore Estate. There is also a modern maze in the grounds of the Breamore Countryside Museum (See page 50).

Wing, Rutland

7. The Wing Maze, Wing, Rutland

This well-kept maze is cut into the verge alongside the road near to the recreation ground. It is of an eleven-ring medieval Christian design. However, a plan of the maze from the last century shows a loop in the middle of the maze which sends the runner back to the start. Later it was changed to conform to the medieval Christian pattern. At one time there was a raised earthen mound on which spectators could stand, but no trace of this now remains.

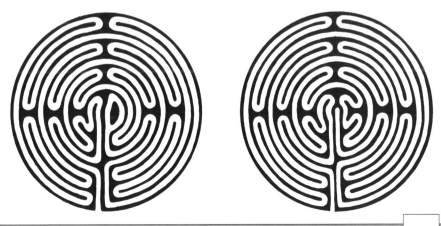

Julian's Bower, Alkborough, Lincolnshire

8. Julian's Bower, Alkborough, North Lincolnshire

This eleven-ring medieval Christian design turf maze is in a spectacular spot on high ground overlooking the point where the River Trent flows into the Humber estuary. From 1080 until 1220 there was a small monastic grange nearby and the making of the maze has been attributed to the monks. One legend says that one of the knights who assassinated Thomas à Becket was given a pilgrimage to Jerusalem as a penance for his crime. He was unable to go on this journey and cut the maze instead. The earliest known documentary evidence for the maze is in the 'Diary' of Abraham de la Pryme written between 1671 and 1704 in which it is called Gillian's Bore.

Whether Bore or Bower there are two different legends which try to explain the origins of the name. Both connect its origin with the River Trent over which it looks. The first says that the spirit of the river, known as Gur, was angry when the maze was cut and was visited by so many Christian pilgrims. He was determined to get rid of it and sent a great wave up the river in an attempt to wash away both the maze and the pilgrims. The wave however did not reach high enough but he continues to try at every spring tide when a small tidal wave known as the Trent Bore rushes up the river past the maze and the village. The second legend associates the maze with the legend of Julian the Hospitator, a nobleman who was fond of hunting. One night when he had not returned from hunting his wife gave their bed to his parents who had arrived unexpectedly. Julian arrived home and finding two people in his bed thought his wife had taken a lover and he slew them both. A fortune-teller had earlier predicted that he would slay his parents and this is how it came about. In order to atone for this crime he set up a hospice by the side of a river where travellers could rest and he ferried them across the river in his boat. One day he transported a leper who was dying from cold and gave him food and shelter in his own bed. In the morning the leper was transformed into an angel who told him that because of his act of generosity he was forgiven his sin.

St. John the Baptist Church, Alkborough, Lincolnshire

Alkborough, Lincolnshire

In the last century the local squire Mr. J. Goulton-Constable took great interest in the maze and re-cut and maintained it at his own expense. When the local church of St. John the Baptist was being restored in 1877 he had the design of the maze cut into the floor of the porch of the church and the grooves filled with cement. The design also appears high on the east window behind the altar.

When he died a Celtic cross, into which has been set a metal replica of the maze, was erected over his grave in the cemetery.

Turf and Stone Designs

Stone Labyrinths

The picture below shows The Troy Town stone labyrinth on the island of St. Agnes in the Scillies. It was constructed in 1729 by the local lighthouse keeper. Over the years its shape has been altered by visitors but it seems likely that its original shape was of the Classical seven-ring design, slightly smaller than it is at present. There is no record of why he constructed the labyrinth and it is the only ancient one so far south. All other examples are in the Nordic countries up to the Arctic Circle and especially on the shores of the Baltic Sea.

St. Agnes, Scillies

There are other stone labyrinths on the neighbouring island of St. Martins but these are of recent origin. The oldest is supposedly a copy of the one at St. Agnes. It is said to have been constructed by an aircrew with time to spare while stationed on the island during the Second World War.

Ancient stone labyrinths in Scandinavia are very difficult to date. Recent work on the lichens on the stones has suggested dates from the 13th century up to the present day. The majority seem to have been constructed in the 16th and 17th centuries. Most are near the sea shore and are associated with fishing communities or burial grounds. Virtually all are of the Classical labyrinth pattern and many have names linking them to the three cities of Troy, Jericho and Jerusalem.

The Trojaborg labyrinth at Visby on the Island of Gotland, Sweden.

The labyrinth on Great Hare Island, Solovecke, Russia.

Fishermen would walk the labyrinth before setting out to sea in the hope of securing favourable winds and a good catch. They were also used by young people in courtship rituals. A girl would stand in the centre and a boy ran the labyrinth in order to win her.

Some of the ones associated with burial grounds may date back to the Bronze Age and almost certainly had a ritual significance. One possibility is that they were used to lure unfriendly spirits into the labyrinth, trapping them there and thus leaving them unable to trouble anyone outside.

This drawing was made by Dr. E. von Baer in 1838 of a stone labyrinth seen by him on the island of Weir in the Gulf of Finland. It is unusual in that it seems to have a short route to the centre and a very long route to a dead end. Could it be that unfriendly spirits only turn left when offered a choice?

Chapter 3
Labyrinths from Roman Mosaic Floors

The earliest mosaic floors were laid in Greece and used small coloured pebbles to form the designs. Sometimes they were outlined in strips of metal. Floor makers then got the idea of cutting marble and other stones into small cubes which could be laid into a mortar on top of a prepared floor to create a variety of designs and patterns. In some cases the floor was made of *opus signinum* into which was laid a pattern of small white cubes. *Opus signinum* was a cement into which was added crushed tiles, bricks and pottery, giving it a distinct brownish-red look. Some floors, especially at Pompeii, had black volcanic ash added to them to give a dark grey-black colour. Two of the surviving Roman labyrinths are made of *opus signinum* - one, a fragment, from Solunto in Sicily and the other from Mieza in Macedonia, Greece.

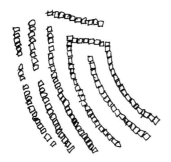

The drawing above is of the opus signinum fragment from Solunto. To the right is an example of another Roman floor from nearby, illustrated here to show the general effect of an opus signinum design.

Solunto, Sicily

While most Roman mosaic floor patterns are not labyrinths, there are around sixty known ones that are. Most are too small to have been walked even by children and since some are positioned at the entrance to rooms it is suggested that they had an apotropaic and protective function. They date from around the second century B.C. to the fifth century A.D. and are found throughout the Roman Empire from Cyprus to Portugal and from England to North Africa. Some have a panel in the centre showing Theseus killing the Minotaur, or sometimes the Minotaur alone, thus linking them with the ancient myth from Crete.

These sixty or so mosaic designs have been the subject of study by many scholars. Dr. David Smith made a list of known examples in an article in November 1959. Professor Wiktor Daszewski published a book containing a catalogue and classification of fifty-four examples in 1977. Hermann Kern added a further five examples in his monumental book on the labyrinth in 1982. Since then articles have been published, notably by John Kraft in 1985 and Anthony Phillips in 1993.

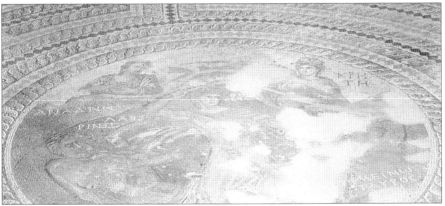

Nea Paphos, Cyprus

This mosaic from The Villa of Theseus at Nea Paphos in Cyprus is perhaps the finest example of a Roman mosaic labyrinth floor. The pathway is patterned with running guilloche ('twisted rope') in blue and pink. The central panel shows Theseus about to strike the Minotaur in the labyrinth. To the left of Theseus is an old man who is the personification of the labyrinth. Above this group to the left is Ariadne and to the right is the personification of Crete. All have their names inscribed in Greek beside them.

Nea Paphos, Cyprus

This small labyrinth mosaic is part of the pattern of a large floor from the House of Dionysius also in Nea Paphos.

Labyrinths from Roman Mosaic Floors

Many of the mosaic labyrinths have a 'fortress' design around them. These are a common feature of Roman mosaics and show the walls of a city or legionary fortress with gateways on the four sides and towers at the four corners. These, perhaps, refer back to the most famous city of ancient times, the City of Troy.

This photograph shows part of the Labyrinth Mosaic from Thurburbo Majus now in the Bardo Museum in Tunis. It depicts one of the corner towers and a portion of the walls of the city. The full plan of the labyrinth is given below.

At its centre is a mosaic of 'Theseus and the Minotaur' which is shown on page 4.

Bardo Museum, Tunis

Avenches, Switzerland

This small circular labyrinth from Avenches in Switzerland is of the 'fortress' style. The labyrinth is surrounded by city walls with four equally spaced gates.

Here are two more examples from Switzerland. One is from from the Villa at Orbe-Bosceaz and the other from Fribourg.

Orbe-Bosceaz, Switzerland

The labyrinth at the villa is surrounded by a representation of city walls. Here is one of the gateways.

Orbe-Bosceaz, Switzerland

The circular labyrinth, drawn above, is from Fribourg. It is unusual because it is divided into eight sectors rather than the customary four. The entrance is via the gateway at the top left corner and all eight sectors must be traversed in turn in order to reach the centre.

Labyrinths from Roman Mosaic Floors

The typical Roman mosaic labyrinth is unicursal. Some examples are circular and some square. They are usually composed of four repeated patterns around a central cross probably suggested by the main roadways in a Roman town or fortress. It is a characteristic that each quarter of the design had to be travelled before starting the next. Finally after the complete labyrinth is traversed, the pathway reaches the centre. It is possible to classify the forms of the pathways traversed into four kinds, Serpentine, Spiral, Simple Meander or Complex Meander.

1. Serpentine
The pathway through each quarter of the design follows a pattern which suggests the way a snake might fold itself into a small space, hence the use of the word serpentine.

2. Spiral
The pathway spirals into the centre of the quarter between alternate pairs of walls and the spirals out again between the others. A complete labyrinth in this design therefore has four false centres to traverse before reaching the true centre.

3. Simple Meander
In this design, each quarter has a series of coils. These examples have three and each coil contains four pathways. Two clockwise to the centre of the coil where it changes direction and two anticlockwise out from the centre to the next coil.

4. Complex Meander
A complex meander has more than two pathways to the centre of each coil. In these two-coil examples there are three pathways. Other examples have four.

Walls of circular design   Walls of square design

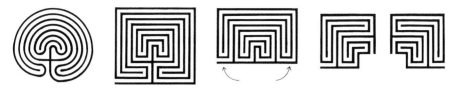

This sequence of drawings shows how the Classical labyrinth can be squared off, opened out and then condensed into a quarter of a standard Roman mosaic design. In fact it can be condensed either way into two quarters which are mirror images of each other. An excavation at Mieza in Macedonia in Greece has uncovered a design which appears to be a link between this form and the usual Roman design with its central cross. Staffan Lundén & John Kraft have analysed the design from excavation photographs.

This is the design as it has been drawn by John Kraft from excavation photographs. There appear to be three loops that are closed off and which are inaccessible from the path through the labyrinth. There are also places where a choice of pathway is offered. It is not certain but it is probable that these are mistakes in the design.

Here John Kraft has 'corrected' the design to make it conform to the requirement of having four symmetrical quadrants. Each of the quadrants derive from a Classical labyrinth adapted in the way that has been demonstrated above.

Later Roman examples have 'corridors' joining up the four quadrants, so forming the distinctive cross. Staffan Lundén suggests that these were added in order to make the labyrinth into one continuous pathway with no choices. They also resemble the roadways of a Roman city or fortress and this may have been a strong reason for adding them.

Conimbriga, near Coimbra, Portugal

This remarkable and well preserved Roman mosaic floor is at the 'House of the Fountains' in Conimbriga near Coimbra in Portugal. It includes two different labyrinths. In the foreground is a two-coil simple meander design reproduced again on page 2. It has a bull's head in the middle, no doubt representing the Minotaur. The adjoining panel has a classical labyrinth design with the pathway laid out in different coloured guillouche. Its plan is given here, rotated by a quarter turn from the way it is usually drawn.

28

In Britain there are six known examples of Roman labyrinth mosaic pavements and a possible seventh.

Caerleon, Gwent

Three of them are almost identical and the example from Caerleon is shown above. They are all of the three-coil simple meander pattern and the other two are in Harpham, East Yorkshire and Oldcotes, Nottinghamshire. The one at Caerleon has a pathway that takes you in a clockwise direction around the central panel whereas Harpham goes anti-clockwise. At Oldcotes, the mosaic was excavated in 1870 and then re-buried. There is no drawing of it but a description of the time said 'The remainder of the design consists of a labyrinth almost identical with that discovered at Caerleon. The centre ..... was unfortunately much injured, but the lower portion of the human figure remained, in an attitude of attack, one arm had been extended with a short broad sword pointed downwards the lower part of the blade remaining; and over the shoulder the outline of an oval shield was evident'. This, surely, must have been a portrayal of Theseus and the Minotaur.

The Harpham mosaic was found in 1904 and re-excavated in the 1950s. It was on display in the City Hall in Hull for some time but recently it has been transferred to the Hull and East Riding Museum and put into store until it can be conserved and put on display again. It has a four-petalled flower in the centre and is unique amongst Roman mosaic labyrinths in that respect.

Patterns from Roman Mosaics

The mosaic below is from a villa in Fullerton, Hampshire. It was excavated in 1872 and again in 1963 and the drawing follows the reconstruction by David Neal.

It seems to be a crudely made labyrinth but follows none of the usual designs. However, it does have a crenellated surround which suggests a fortress. Perhaps the mosaicist was out of his depth in trying to copy a design that he did not properly understand.

David Neal has also reconstructed this Roman mosaic labyrinth discovered in Cirencester.

It is of a two-coil simple meander design but the four corners have been indented to make it into the shape of a cross. This treatment has the effect of making the design visually much more exciting. The zig-zag effect makes it particularly attractive and dramatic and yet retains the single pathway leading to the centre.

This Roman labyrinth design from another building in Cirencester provides a border to a central square mosaic. The reconstruction drawing is by Stephen Cosh.

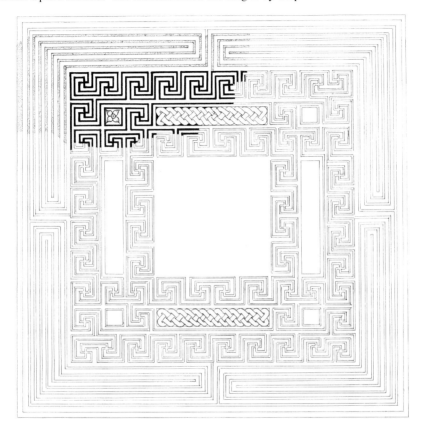

This drawing of a damaged floor in a corner of a room of the villa at Keynsham shows what might be a possible seventh Roman labyrinth design. It has been taken from a painting by Stephen Cosh. As it cannot be reconstructed to any of the usual meander designs he thinks that it could be a labyrinth. However it cannot be reconstructed to conform to any of the usual mosaic labyrinth patterns either. Unfortunately the remainder of the room is buried under a later cemetery chapel and the problem cannot be resolved until the site is cleared and the mosaic re-excavated.

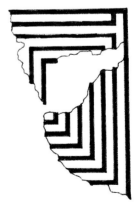

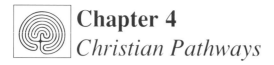

Chapter 4
Christian Pathways

There is a long history of the use of labyrinth designs in churches. However, the reason for their presence and their exact historical significance is not known. It is thought that they were used as a form of penance with the penitents 'walking' the maze on their knees. Possibly they were a substitute for going on a pilgrimage or navigated as a spiritual exercise in prayer and meditation. Indeed, the pathway of the labyrinth does resemble the tortuous passage through life with wild swings towards the centre only to return to the outside before achieving the final goal. Many labyrinths have been laid in modern churches, especially in America, with just such a purpose in mind and the idea of the labyrinth as a spiritual exercise has certainly grown in recent years.

The oldest known labyrinth in a Christian Church is in the Church of Reparatus, at El Asnam, in Algeria. It is dated 324 A.D. and is made of mosaic. It follows the Roman spiral pattern.

A	I	S	E	L	C	E	C	L	E	S	I	A
I	S	E	L	C	E	A	E	C	L	E	S	I
S	E	L	C	E	A	T	A	E	C	L	E	S
E	L	C	E	A	T	C	T	A	E	C	L	E
L	C	E	A	T	C	N	C	T	A	E	C	L
C	E	A	T	C	N	A	N	C	T	A	E	C
E	A	T	C	N	A	S	A	N	C	T	A	E
C	E	A	T	C	N	A	N	C	T	A	E	C
L	C	E	A	T	C	N	C	T	A	E	C	L
E	L	C	E	A	T	C	T	A	E	C	L	E
S	E	L	C	E	A	T	A	E	C	L	E	S
I	S	E	L	C	E	A	E	C	L	E	S	I
A	I	S	E	L	C	E	C	L	E	S	I	A

At its centre is a panel on which there is a cleverly arranged square of the letters of the Latin words for 'Holy Church'. At the centre of this panel is the letter S and from it the words 'Sancta Eclesia' can be read in any combination of horizontal or vertical directions. The 'a' of 'Eclesia' always ends in one of the four corners.

Another early but undated labyrinth is carved into a column in front of the cathedral in Lucca in Italy. A translation of the Latin inscription at the side reads:

'This labyrinth is the one that Daedalus the Cretan built. No-one who entered it could get out except Theseus and he could not have done it without being helped by Ariadne with a thread and her love.'

Lucca, Italy

In the sixth century church of San Vitale in Ravenna, among all the wonderfully patterned floors of coloured marble, is the labyrinth in the picture below. Doubt exists as to whether it is part of the original Byzantine floor or whether it belongs to the sixteenth century re-construction of the floor after it was damaged by floods. However, even if it does belong to this later period it could still be a copy of the original. This type of Christian labyrinth is usually found from the eleventh and twelfth centuries in churches in Italy and France. If it does belong to the sixth century then it is very early indeed and no other similar labyrinths have been found in the Byzantine Empire. The pathway is of light coloured triangles laid into a darker surround. The triangles point from the centre outwards.

Ravenna, Italy

The high point for the labyrinth as a design for the floor in churches was during the eleventh and twelfth centuries, especially in France. As has been stated on page 9, perhaps the most famous example of such a labyrinth is that which was laid in 1202 in Chartres Cathedral. It is made of blue and white stone and is 12.8 metres in diameter.

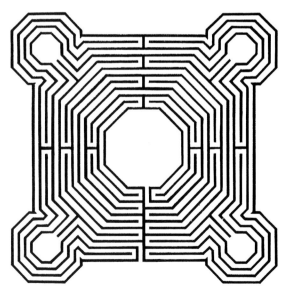

This magnificent example was once in Rheims Cathedral and was laid early in the thirteenth century. It was destroyed in 1779 by Canon Jacquemart who objected to the noise of children walking the maze during services.

Here are some more labyrinth designs from France and both of these have survived to the present day.

This circular labyrinth is made of red and black tiles and encaustic tiles, patterned in brown and yellow with shields, fleur-de-lys and griffins. It can be seen in the chapter house of Bayeux Cathedral, not far from the Bayeux Tapestry.

This octagonal design comes from the parish church of St Quentin in Northern France.

This square pavement was once in the Abbey of St. Bertin in St. Omer in Northern France. It was destroyed in the eighteenth century but its design survived. A copy can now be seen in the parish church of St. John the Baptist at Batheaston near Bath.

Batheaston, near Bath

Bourn, Cambridgeshire

This picture shows a maze laid in black and red tiles which is of a design that did not come from France. It is based on the hedge maze at Hampton Court but with its lines straightened and the curves squared off. The font stands at the centre of the maze and it lies at the west end of the church of St. Helena and St. Mary in the village of Bourn just to the west of Cambridge. It was made in 1875.

Bourn, Cambridgeshire

Christian Pathways

Both of the labyrinths on this page use the famous Chartres design.

Itchen Stoke, Hampshire

St. Mary's Church at Itchen Stoke was designed and built in 1866 by Henry Conybeare who was the brother of the vicar at the time. He based the overall design on the beautiful La Sainte Chapelle in Paris. This copy of the Chartres labyrinth is made from green and brown glazed tiles and is in the chancel.

Royston, Cambridgeshire

This is an example of the Chartres design being used in a non-religious context. It was made in glazed tiles by Maggie Berkowitz (See page 64) for a company headquarters.

Earlier in the book there were several labyrinths from Roman Switzerland. These two are beneath the signs of the four Evangelists on the modern bronze doors of the church of St. Francis in Lausanne. They are unusual because they show the pathways through the labyrinths and not the labyrinths themselves. One is a simple spiral and the other is the Classical seven-ring design.

Lausanne, Switzerland

Lausanne, Switzerland

Lausanne, Switzerland

Ely, Cambridgeshire

When Ely Cathedral was being restored by Sir Gilbert Scott in 1870 he placed this black and white stone labyrinth with four symmetrical bastions just inside the west doors and directly under the west tower. In a leap of imagination he had it constructed so that the length of the pathway through the labyrinth was exactly the same as the height of the tower immediately above. Both are 215 feet.

The late Victorian period was a time of great self-confidence and relative prosperity. Apart from much restoration work on older churches and cathedrals, many new churches and chapels were also built. It was a time when the imaginative use of bricks and tiles for decoration reached new heights.

Just outside the village of Compton in Surrey stands a beautiful Romanesque-style red brick and terracotta chapel. It was designed and decorated by Mary Seton Watts as a funerary chapel for her husband, the painter George Frederick Watts, who died in 1904. It is a triumphant mixture of Art Nouveau and the Arts & Crafts movement.

Compton, Surrey

Compton, Surrey

Compton, Surrey

There are two labyrinths included in its design, one on the outside and one inside. Outside, four groups of three angels act as corbels and one in each set of three holds a labyrinth which is a copy of the one at Ravenna, described on page 33. Inside is an altar made of gilded terracotta. One of the panels on it includes another angel, this time holding a labyrinth to a similar design but extended to encompass eleven rings.

With this impressive artistic achievement, drawing inspiration from the past, we bring this small selection of labyrinth designs in churches to an end. There are many more to find.

Chapter 5
Hedge Mazes

In Europe during the Middle Ages formal gardens began to be laid out using walls or hedges to keep out foraging animals. Herb gardens soon proved popular and began to be laid out in symmetrical geometrical patterns. At first some herbs were clipped into low hedges but later box hedges began to be used to edge the beds. Gardens like these became known as Knot Gardens and sometimes traditional labyrinth designs were used. Such labyrinths offered a single pathway to the centre. Eventually, someone had the idea to use yew or other taller plants instead of box for the hedges with the result that the centre could no longer be seen. It is only a short step from here to offer a choice of pathways and the opportunity of getting lost, thus changing the unicursal labyrinth into a multicursal maze. Such hedge mazes became popular initially in Italy, Germany, France and Holland and then rather later in Britain.

Pitmedden, Aberdeenshire

Above is 'The Great Garden' at Pitmedden in Aberdeenshire. It is a modern reconstruction of a garden first laid out in 1675. To the right is the walled garden at Edzell Castle in Angus.

Edzell Castle, Angus

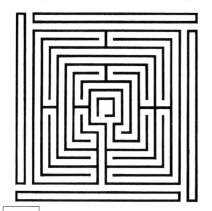

This is the plan of the hedge maze planted by Lord Burleigh at Theobalds in around 1560. Later, his son the Earl of Salisbury, exchanged Theobalds for Hatfield House and the new owner was King James I. Theobalds became his favourite residence. The design of the maze survives, but not the maze itself or the house. Both were destroyed during the Civil War in 1643.

The oldest remaining, and certainly the most famous, hedge maze in England is the one at Hampton Court. It was planted in 1690 for William of Orange when he and Queen Mary remodelled the grounds after coming to the throne in 1688. William had earlier had constructed a hedge maze at his palace at Het Loo in Holland.

Hampton Court Maze

Many hedge mazes and formal gardens were destroyed in England in the eighteenth century when 'landscape gardening' became the craze. However, in the following century formal gardens once again became popular and new hedge mazes were created. One in particular, at Chevening in Kent, was made in the 1820's using a design by the second Earl Stanhope who was a mathematician. He had realised that any maze which had its hedge as a single entity, no matter how branched, could always be solved by the 'hand-on-hedge' method. By walking so that the face of the hedge is consistently on the right, or on the left, a maze-walker can always arrive at the centre or find the way out. It is not necessarily the shortest route but it will always solve the puzzle. However, if the hedge is separated into several parts or 'islands' it cannot be solved in this way. In fact it can be very difficult to solve at all. The basis of the design is a mixture of the Roman mosaic and Christian labyrinths with parts removed so that islands are created. This maze is still at Chevening but is not open to the public.

This shows how the maze is made up of several islands of hedges. The goal is in a separate island from the perimeter.

Hedge Mazes

The design of the maze at Chevening is about as difficult as a puzzle maze can get without the addition of a third dimension in the form of bridges or tunnels. Some modern mazes use these techniques and in 1978 a hedge maze with six bridges, designed by Greg Bright, was constructed in the grounds of Longleat House in Wiltshire. At present it is the largest hedge maze in the world. Since then two more mazes designed by Randoll Coate have been planted at Longleat using the sun and moon as themes. Further mazes are planned there in the near future.

Another development in traditional hedge mazes can be seen at the Dragonfly Maze in Bourton-on-the-Water in the Cotswolds. Planted in 1996, it includes a puzzle which has to be solved from clues gathered on the way through the maze. Kit Williams, the author of 'Masquerade', designed both the rebus puzzle and the centre-piece.

Scone Palace, Perthshire

This beech hedge maze at Scone Palace near Perth, Scotland was designed by Adrian Fisher. It is in the shape of the five-pointed star of the Murray family and planted with green and copper beech to give the appearance of tartan.

As printing became more affordable, many books on the art of gardening and of architecture were published. Many had diagrams of mazes and three early designs using the same basic shape are shown above. Two are from 'Horicultura' published in 1632 in Frankfurt and the other comes from a French book of the same period.

These two designs were published by André Mollet in 'Le Jardin de Plaisir' in 1651.

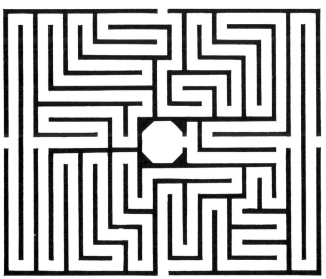

45

The designs on these two pages are by G. A. Boeckler in a book 'Architectura Curiosa Nova' published in 1664.

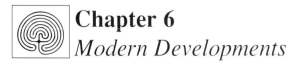

Chapter 6
Modern Developments

The remaining pages of this book are devoted to some of the hundreds of new mazes that have been constructed in recent years. Maze design is an exciting modern field of considerable creativity, imagination and ingenuity. However, while the designs themselves are often completely original, it is interesting to see that the designers have not forgotten or ignored the long tradition that preceded them.

Warren Street Underground Station, London

Warren Street in London was built on land belonging to Lord Southampton, who named it after his wife Anne Warren. In the late 1960's/early 1970's when the new Victoria Line was built, it was decided that all the new stations should be decorated with ceramic tiles and that each should have an appropriate theme. The choice for Warren Street was based on a simple pun (warren = maze) and these striking panels in red, black and white showing a maze in perspective were the happy result.

In 1979, a pavement maze, designed by John Burrell, was laid in a children's playground not far from the Underground station, so that pun has continued to maintain its power and influence. Like the design at the station, this is a true maze not a labyrinth, with choices, dead ends and repetitive loops. However, trace it and observe that the shortest pathway to the centre is a long one, reminiscent of traditional labyrinths.

Warren Street, London

Modern Developments

In 1984 The Sunday Times in association with WATCH organised 'The Great British Maze' competition. Ian Leitch of Aberdeen submitted the winning entry and his maze was constructed at Breamore House in Hampshire not far from the Breamore Mizmaze described on page 16. In the terms of the competition hedges were not allowed and the maze had to be constructed of brick pathways in turf. He took as his theme the shape of a five-barred gate and, in an imaginative masterstroke, proposed a yew ewe to be grown at its centre.

Breamore, Hampshire

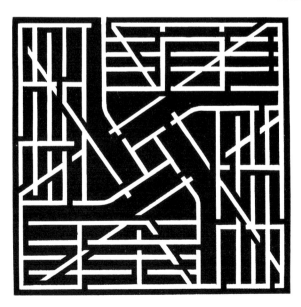

King Edward Street, Hull

When King Edward Street in the centre of Hull was to be pedestrianised in 1987 ideas were sought for what might be included. Philip Heselton suggested a brick maze be built and he was asked by the City Council to design one. Looking for a suitable theme he remembered that the historic Julian's Bower at Alkborough was not far away and based his design on it. His is a squared-off version and he reduced the eleven-ring medieval Christian labyrinth to nine rings but the local historical connection is firmly there. In the centre is a plaque which draws attention to the fact that this maze has a single path to the centre and no dead ends and that in medieval times such mazes were often run or danced on festive occasions.

Willen Park, Milton Keynes

In Willen Park in Milton Keynes, near to the lake and close to the Peace Pagoda, there is a maze which has a pathway which is over two miles long. It was made in 1984 and is based on the design of the Saffron Walden Town Maze, described on page 13. However, additional pathways have been cut across in order to change it from a simple labyrinth into a puzzle maze. In the four bastions there are four bronze faces designed by Tim Minett and there is a tree at the centre.

Crawley, Sussex

In the County Mall in Crawley, there is a very interesting double puzzle pavement tile maze which was designed by Adrian Fisher in 1992. It was inspired by the work of the artist M.C. Escher and can be entered from either side. It is distinctive in that a white pathway with black walls is at the same time a black pathway with white walls. Look in particular at the pathways near to the centre to see how this works. In the centre is a mosaic made by Emma Biggs. Within the same County Mall is another colour maze by Adrian Fisher which represents a game of hopscotch.

Crawley, Sussex

Victoria Park, Bristol

There is a very interesting water maze in Victoria Park in Bristol. It was designed by Peter Milner and Jane Norbury in 1984 to commemorate the opening of a new sewerage system. This suggested a design based on running water and they conceived the idea of the pathway through a maze as a channel of flowing water. The complete maze is built of brick and the water flows from the centre and follows the pathway to the outside. This required a unicursal labyrinth design and by good fortune there is an eleven-ring medieval Christian labyrinth on a carved and gilded roof boss in the church of St. Mary Redcliffe nearby. The spire of this church can be seen from Victoria Park and the maze is aligned to it. This is a fine example of the way traditional and modern ideas are brought together in modern maze design.

St. Mary Redcliffe, Bristol

In Bath there is an elliptical maze created in 1984 by Adrian Fisher with a central mosaic designed by Randoll Coate. In that year the theme of the Bath Festival of the Arts was 'The Maze' and a major exhibition of the work of Michael Ayrton was held. At the same time, Michael Tippett's opera 'The Knot Garden' was performed and the decision was made to construct a permanent maze in Beazer Gardens near to Pulteney Bridge. It is known as the Bath Festival Maze.

Its elliptical form is unusual and Adrian Fisher took his inspiration from three special features of Bath; the curve of the weir that the site overlooks, shapes of fanlights in the Georgian houses and the arches of Brunel's famous railway. The circular centre is a mosaic which looks to more traditional roots with six semi-circular apses recalling the centre of the labyrinth at Chartres. The central panel is a powerful image developed from a Gorgon's head which was found on a pediment of the nearby Roman temple. It is surrounded by six semi-circular images combining myths from Bath's Roman and Celtic past. These mosaics are also 'Gaze-Mazes' linked by a 'thread' of coloured marble. It is intended that the eye should follow this thread through the design.

Gorgon's head, Bath   Beazer Gardens, Bath

Modern Developments

Elaine M. Goodwin is an internationally known artist and the labyrinth has been one of the recurring themes and inspirations for her mosaics. On these two pages are five of her designs which include labyrinths combined with mythological images and ideas. Together with Group Five, she has created a series of fourteen large murals for schools, car parks and public places in Exeter. This work evolved from a six-month visit to India in 1983-84 when she worked with Nek Chand on the twenty-five acre Rock Garden at Chandigarh. This garden includes many examples of mosaics of life-sized figures, animals and mythological scenes which have been made from broken crockery and other scavenged material.

In this imaginative design, elephants are shown tracing out the path of a labyrinth which leads to a rat at the centre. Its inspiration is the elephant-headed god Ganesh.

Among Hindus, Ganesh is worshipped as a remover of obstacles and as a patron of learning and is often portrayed riding upon a rat.

Ganesh goes Walkabout

The Chase

This design shows a circle of animals surrounding a circular labyrinth with a human figure at its centre.

This represents humanity's crucial role in the evolution of our planet and in protecting the animals that live with us. The circle of animals is bordered by four smaller corner labyrinths.

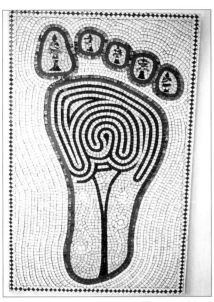

Giant Foot Maze

In Indian tradition a footprint brings good fortune and a labyrinth offers protection and so this design is very appropriate as a welcoming panel to a house.

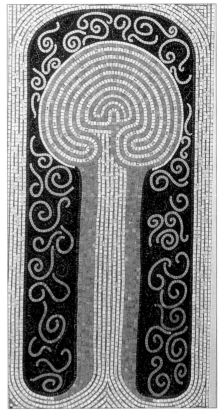

Tree Maze

This design presents a tree of life with the journey from birth to death depicted as a pathway through a labyrinth.

Here seven lizards surround a labyrinth based on the one from San Vitale in Ravenna which is described on page 33.

Modern Developments

Adrian Fisher has been a great influence on modern maze design and has been at the forefront of the explosion of maze building which has taken place in the past few years. His first hedge maze was constructed in 1975 and since then he has created more than 175 mazes all over the world. Three have already appeared in this book. They are: the hedge maze at Scone Palace on page 44, the maze at County Mall at Crawley on page 53 and the Bath Festival Maze on page 55. He searches for new themes and new possibilities and is keen to work with new materials. His mazes have been created from hedges, wooden fencing panels, mirrors, brick pavements, terrazzo and tiles, mosaic, turf, water, plastic and maize. They are sited in country houses, parks and public places, schools and zoos. Some, like those at Legoland, Windsor and the maize mazes attract many thousands of visitors each year.

Norton Museum, Florida

In 1977 he had a one man exhibition 'An Amazing Art: Contemporary Labyrinths by Adrian Fisher' in the Norton Museum of Art, West Palm Beach, Florida. For this exhibition he created a permanent brick pavement maze portraying 'Theseus Slaying the Minotaur' and a wall tile mural 'Double Spiral Maze'.

Double Spiral Maze, Norton Museum, Florida

Russsborough House, County Wicklow

In Russborough House at Blessington in County Wicklow, Ireland, he together with Randoll Coate designed this formal maze which has its centre in the form of a diamond. This shape was suggested by the fact that the owner's family were pioneers in the diamond industry. Its centre is marked by a statue of Cupid standing on a column and high enough to be seen from all parts of the maze.

Mary Hare Grammar School, Newbury, Berkshire

The Mary Hare Grammar School is a school for the deaf and the maze has a giant seashell as a pun on the labyrinth of the inner ear. It is in the Sixth Form courtyard.

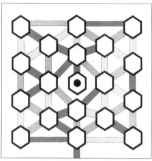

In 1986, Adrian Fisher conceived the idea of a new kind of maze with pathways in a number of different colours. At each junction or decision point, the maze runner had to obey a simple rule to determine which pathways were permitted and which were forbidden. The junctions were called 'nodes'.

Such mazes draw on elements of mathematical topology and require logical thinking and persistence to find a solution. This innovative idea opened up a considerable change from the processes of traditional mazes and in December 1986 he wrote an article about them for 'Scientific American'. For the 'Mathematica Colour Maze' the rule says that the runner should change path colour in the sequence red/blue/yellow each time a node is reached. It starts on the first red path and, as in conventional mazes, the aim is to reach the centre. It is additionally ingenious in that a second sequence of red/yellow/blue produces a different solution.

Here are the two solutions.

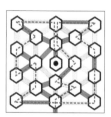
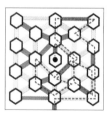

Leicester University

The 'Mathematica Colour Maze' which was constructed outside the Mathematics Building of Leicester University in 1991 is topologically identical to the colour maze illustrated above, but the hexagons have become squares and the colours have been changed to shades more suitable for brickwork. In addition, each of the square nodes is labelled with a letter of the alphabet and one sequence will spell the word 'Mathematica'.

In 1993 Adrian Fisher designed the first ever maize maze with the pathways cut from a field of growing corn. The theme for the design was a Stegosaurus dinosaur. Covering 126,000 square feet it had nearly two miles of paths and featured in the Guinness Book of Records as the world's largest maze. This record has since been broken by himself with successive maize mazes in 1995, 1996 and 1997 as the idea proved a very popular one and attracted many thousands of visitors.

Lebanon Valley College, Annville, Pennsylvania

In 1998 he designed 12 and in 1999 a total of 24 maize mazes at locations across the world. Nine of the 1998 mazes were on the theme of dragons and this included the three sites in England. On Millets Farm the subject was Mark the Mythical Dragon.

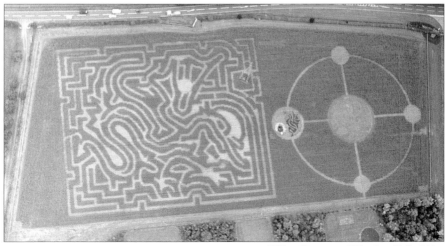

Millets Farm near Abingdon, Oxfordshire

Modern Developments

Michael Ayrton was a painter, sculptor, theatrical designer, book illustrator and author. In May 1956 while on holiday in Italy he went to the point at Procida, west of Naples and looked down the long isthmus towards the great rock at Cumae - the traditional spot where Daedalus came down to earth after his long flight - and a creative obsession with the myth of Daedalus the mazemaker took him over. The result was many years work in drawing, etching, painting, sculpture and writing. Some of his finest work was executed in response to this creative impulse.

Among these are: a series of etchings on the theme of the Minotaur, paintings and drawings of Cumae, many bronzes of Daedalus and Icarus, the Minotaur and an incredible series of maze heads. He also wrote an 'autobiography' of Daedalus called 'The Maze Maker', described by his publishers as a novel and succeeded in casting a honeycomb in gold. In this way he copied one of the feats of Daedalus himself.

Dust Jacket of 'The Maze Maker'

End Maze III, University of Exeter

The Maze at Arkville

Armand G. Erpf, a New York financier read 'The Maze Maker' when it was published in America in 1967 and asked Michael Ayrton to build him a maze on his estate at Arkville in the Catskill mountains.

This remarkable maze is built of stone and brick and contains an estimated 210,000 bricks. The walls are eight feet high and it has two central chambers. One contains a dramatic bronze of the Minotaur and the other a bronze of Daedalus at work on the maze while Icarus leaps upward from his shoulders to try out his wings. The chamber of Daedalus and Icarus is lined with bronze mirrors.

Postmen's Park, London

Hadstock, Essex

A copy of the Arkville Minotaur now stands in the City of London and a bronze replica of the Arkville maze is on the back of the headstone of the grave of Michael Ayrton and his wife Elizabeth at the village of Hadstock near Saffron Walden in Essex.

Maggie Berkowitz is an artist living in Cumbria and working mainly in ceramics. On page 38 is one of her tiled floors based on the Chartres labyrinth and below is a fine example of a pictorial maze created for the entrance to a house in France.

A private house in France

Low Furness, Cumbria

Low Furness, Cumbria

She also designed this 'Journey to Jerusalem' labyrinth for Low Furness Church of England Primary School, Cumbria, including as a goal a central tiled panel which represents the ancient city of Jerusalem.

With this last labyrinth a full circle in the exploration of the history and development of labyrinths and mazes has been completed. Starting with the story of Theseus and the Minotaur and the labyrinth as a symbol of ancient cities and mysteries the journey has ended with ingenious modern interpretations and developments of these same themes. Despite changing fashions and ideas the maze and the labyrinth have been able to offer fertile and creative inspirations for artists and designers through the ages. There is no reason to suppose that this will not continue and we await the future with interest.